LADY
GAIA
SPEAKS

Patricia Finney

First published 2018
Copyright © 2018 Patricia Finney

ISBN 978-1-909172-39-5

More books by Patricia Finney may be found at
www.climbingtreebooks.com

Published by Climbing Tree Books Limited,
Truro, Cornwall, UK

Typeset by Grace Kennard, Penryn, Cornwall

She's about 3.5 billion years old. She extends deep into the rocks of the planet and perhaps a mile or three up into the stratosphere where bacterial spores float. She is not a cuddly little old lady, nor is she a beautiful and kindly woman. She is mostly bacterial and always has been, formed of a complex mixture of prions, viruses, bacteria, algae, fungi, prokaryotes, eukaryotes and – over only the past 350 million years, the last tenth of her life – a tiny percentage of multicellular creatures, from the tiny hydra to the blue whale, from the tiny filaments of slime moulds to the red sequoias.

In about 2 billion years from now she will die as the sun switches to fusing helium not hydrogen and starts to swell. Soon even the deep-sea chemo-eating bacteria will die as her seas boil into space. Perhaps deep in her rocks, some bacteria will survive.

Some time after that the sun will simply swallow her planet. Everything is temporary.

We humans have been here for a tiny sliver of time – as a species for perhaps 500,000 years on current knowledge. In the last 200 years we have started radically to alter her atmospheric chemistry, taking her rapidly back to a time when there were no ice caps and the whole world was hot. Everything is temporary, especially climate.

Our arrogance knows no bounds. We're afraid that we might kill her. It doesn't seem to occur to us that she's survived a lot worse in the last 350 million years since multicellular life started disrupting her calm bacterial equilibrium. Think of the Permian Great Dying when 95% of species died, or the famous asteroid strike 65 million years ago.

Species come and go, she might reassure us. I will live as long as the sun behaves himself.

What else might she say to us? In my earlier book,

Arguments with Our Lady Gaia [Climbing Tree Books, 2013; published as an ebook under the pseudonym Rose Wagner], I've imagined an enormous ancient planetary organism, as in James Lovelock's marvellous books, and imagined what she might say to our teeming billions. Maybe that is what I'm doing as I set out here to write whatever Our Lady Gaia wants.

Or perhaps I'm channelling something, doing the woo-woo thing that claims direct contact with angels and guides and alien species. If them, why not the planetary organism herself? Don't we all carry trillions of her bacteria in our guts and on our skin? Why shouldn't we be able to talk to the vast being of which we're all part?

Maybe I am, I don't honestly know. Of course, just because I believe I can talk to the planetary organism, doesn't necessarily mean that I can. Stuff comes into my head and I write it down - I don't know where it comes from.

Does this matter? No, you can assume that this is just a work of semi-fiction or that I am a whacko nutcase. I don't care, so long as you read this.

Here is Our Lady Gaia, our planetary organism, in all her magnificence, complexity and beauty. She's speaking to us.

What's she saying?

YOU FUCKING MORONS!

You've always been a bit silly, a bit flaky. After all, you're fundamentally monkeys, you don't need the focus of a predator or the patience of a ruminant. You've got constant curiosity and an urge to solve problems, always on the look-out for new and exciting things, new colours and shapes, new food. You started as prey animals, your young regularly eaten by leopards. In a similar way to meerkats you learned to dance and sing to scare dimmer predators off their kills. You became occasional predators, true omnivores, and some of you specialised in predation.

You think you're so wonderful, balancing proudly on your hindlegs, waving the grasping hands you think are so important although many many other animals have had them. You think your colour vision is great – it's poor. Birds and bees have a whole fourth primary colour, not just three. You're able to hear quite well, your sense of smell is good though you ignore it.

You morons! Don't you know how dangerous your position is? Don't you realise the expense and difficulty of bringing up a technological civilization like yours? Look around the stars. There aren't many other civilizations rushing to contact you, are there? Why? Because star-faring civilizations are rare, spatially and temporally.

All right. I understand you simply don't get why I've been so patient with you, despite your ignorance and vandalism. After all, you ludicrous monkeys still think you're the crown of creation, top mammal.

You're not. I'm sorry to have to tell you this but you're my larvae, my spores. Slime moulds make spores and so do I. You are the spores. Your mission (whether you choose to accept it or not) is to use your technology to get off my chunk of rock, create sustainable space habitats and colonise other planets. That's your species mission. You each of you individually have a spiritual mission, but I'm not interested in them. I too have a spiritual mission, but frankly that's my business.

Remember this. As a species your purpose is to move DNA/RNA life off me and ultimately onto other planets going round other suns. That's why I've been so patient with you. As long as there's some chance you'll get cracking on your ONLY JOB, I will tolerate your bone-headed stupidity. But you're taking your sweet time about it and at the moment your stupidity means you don't have much time.

My oceans have started to die. Your greedy extraction of fish and krill in giant factory ships is killing them. That's why there are dead zones in them, and Sargasso Seas full of non-rotting plastic.

The worst-case scenario is that the blue-green algae in the sea which give you your oxygen will begin to starve to death all over the world, because the warm oceans have fewer cold upwellings to keep nutrients at the surface. You won't kill me, because you'll die in all your billions when the blue-green algae die because there won't be enough oxygen in the atmosphere.

The last multibillionaire will hunch over his last oxygen cylinder on his fancy island, and wonder where it all went wrong, since he doesn't believe in global climate change.

If the seas die, you die. I proved this in the Permian Great Dying. 95% of all species disappeared and my dark sister, the part of me that breathes methane and feeds on sulfides, nearly took over. I didn't die. Give me a sea and some algae, and I'll rise again, methane is less reactive than oxygen. Everything else did, very nearly. It was boring and lonely for a while, because I love variety and change.

There are so many ways you can kill yourselves off and you're playing with all of them. Monkeys, admiring the shiny sword in the jungle, the sleek machine gun hanging on a branch.

A nuclear winter would probably lower my temperature enough to trigger an ice age, but ten years without a summer will be enough to wipe out your civilization. You won't get the benefit of the glaciers expanding again. Maybe there'll be a few fur-clad barbarians wandering around the tropics, fighting each other, for a few generations until the last of them dies. The dolphins and whales will rejoice and sing daylong complex epic songs about the mysterious stone whales that killed so many of them. You won't be there.

A plague would do it too – man-made or natural, created by me. I think it would cause much less damage except where nuclear power stations melted down or nuclear bombs were triggered in a stupid attempt to sterilise certain areas – trust me, that wouldn't work. I heal fast, as you've seen with Chernobyl. Take away you greedy monkeys and in 30-50 years I'm beautiful again.

A plague would be my choice, if I decide I've wasted my time because with high enough mortality, your civilisation would fall and you with it. I'm tempted, really I am. It's just that I've spent 100,000 years nurturing you and while that's a very short time, of course, I've also allowed you to wipe out a lot of beautiful species and destroy a lot of habitats, and I don't want to waste the effort.

Yes, I always lose species when I spawn - you need such a high proportion of the spawning species in the biosphere to build a technological civilization.

Why do I want technology? Obviously, because you can't physically cross space without high technology. That's non-negotiable. If I want my DNA/RNA to travel to different planets and grow there, I need radiation-proof physical containers with some kind of propulsion system to get them there. Yes, bacteria can ride on rocks and dead space ships, but generally enough don't survive to colonise a planet and their DNA/RNA is often damaged. Panspermia is quite hard and takes a lot of technology, funnily enough.

You can't sustain technology with fewer than three billion of the spawning species – six to seven billion is better. So you lose species. Over 8 billion, of course, and the whole thing starts to fall apart, especially if the spawning species is as dirty and careless as you are.

Me *(interrupting).* You talk about other Earth spawning species in the past in *Arguments with Our Lady Gaia*, but there isn't much evidence of them. Why aren't there roads visible all over the moon and visible installations? Why aren't we surrounded by the archaeology of the spawning species that went before us, especially if there were so many of them? I understand why there aren't traces on earth – plate tectonics and erosion would hide them, but what about in space?

Lady Gaia. This is the problem with a foraging migrating species. You did learn to bury your shit, but you litter everywhere. It's taken you thousands of years for any of you to get the idea that it's not clever to foul your own nest. And most of you still haven't got it.

The caterpillars were obsessive tidiers. The protomammals were cleaner than you. The dinosaurs learnt to clear up after themselves.

Me. Hm. Convenient. I'd really like to see some actual physical evidence of those wonderful spacefaring dinosaurs you talked about last time. I'll settle for some exciting foot prints on the moon.

LG. Coming up, though you won't interpret it correctly. Also look in what you call the junk DNA. Everyone puts their marker there.

Me. My lady, what is the point of this? We covered it in the last book - which hasn't sold at all well, so nobody's seen it. You can rant and rail about us all you like, but it's not going to do any good without some serious undeniable miracles – and I don't mean hurricanes. You tried that.

LG. Yes, explain to me why the multiple disasters I've given the Americans haven't worked? I even helped them elect a truly appalling President, to give some traction, some opposition to fight against. That civilized and well-meaning President they had earlier just let them slide deeper into smugness and apathy.

Me. I'm sure you've heard of fight/flight/freeze. We're particularly good at freezing and hoping the predator doesn't notice us, since we're not that great at fighting without weapons. Large numbers of Americans are just frozen in fear at what they've done in the last 70 years to harm you. That always leads to loud, in fact hysterical, denial and a refusal to see anything except comforting lies. They don't want to see global warming because if they can see it, they'll have to start fixing it and they're terrified at the lifestyle change involved. They also don't want to see global weirding because they know who largely caused it and they feel guilty.

LG. What absolute stupidity! I know that the tribe of Americans is ridiculously attached to their lifestyle, when they would have much more fun if they just forgot about that. They were the first tribe to escape the Earth, the first to go to the Moon. They will always have a break from me because of that, though if none of you follow through, they'll die like the rest of you.

I've seen to it that the technology lead has passed to the tribe of Chinese who are very capable, although they don't seem to have the combination of wild ingenuity and risk-taking along with focussed detailed engineering smarts of the old Americans. Even that has gone due to so many of their ablest young going into finance. As you know I love capitalism, hate finance.

Me. Why? Look at the way end stage capitalism is hugely increasing the wealth gap between rich and poor.

LG. I don't care about that. Capitalism, when it's

working properly, reproduces a rich ecology in the metaworld of your culture. I've been studying the fantasy you call economics.

Me. You have?

LG. Yes. Do you think I'm stupid and can't learn?

Me. Frankly yes. Didn't you say you're bacterial?

LG. Certainly I am, but bacteria in the mass can make a kind of intelligence through their electrical contacts. When I have an intelligent species in me, I become self-conscious through them, partially through my microbes in their guts. The species just needs a complex brain sufficient for full self-conscious soul-signal reception. And then you need enough of them – four or five million at least.

For instance I've been conscious through the whales and dolphins for 20 million years, skipping species as they go extinct or lose intelligence. It's hard to stay smart, you know. Complex brains are energetically expensive and there is strong evolutionary pressure for lowering the average.

I've been conscious through you monkeys...

Me. Excuse me, but aren't we primates, apes?

LG. Yes, you are but ancestrally you're monkeys and you behave more like them than, say, gorillas.

Me. Gorillas are herbivores, they're serious. What about bonobos? They're nice?

LG. If that's what you call constant fucking, sure. Also calling you monkeys makes me more patient with your antics.

Me. Well, good.

LG. So as I was saying, I've been conscious through you bonoboid monkeys for about fifty thousand years, of course inspiring all the Mother Goddess cults, later Mariolatry, the cult of Kali and so on. That bloodiness of Mother Goddesses happened because being originally prey animals that frequently got eaten, you confuse what you call gods and goddesses with your ancient predators.

There were whole cultures in prehistory that tamed cave-lions by feeding them part of the kill so they would let you live in the mouths of their cave and light nice warm fires there. It was quite a precarious existence but all the humans had toxoplasmosis so they weren't so afraid of the lions. And of course, the parasite made their menfolk slightly crazy so they tended to do well in battle, and that culture was very successful in Ice Age Europe, near the glaciers.

The idea of a god as a meat-eating creature that gave favours and success in battle took hold and became general. Sarah Blaffer Hrdy has this very well.

Me. Wow! Fascinating. Any actual evidence for this? I've read her book so this could be from me.

LG. It could. I don't know, why don't you look? So

you try and feed meat to the gods and even your own bodies. Utterly wasteful and daft. Still that seems to have improved. At least some of you only offer God symbolic meat and blood now.

Me. You're referring to the Christian eucharist?

LG. And the incense, flowers and cakes in Hinduism and Buddhism. Why does a god need to eat anything? It's silly. I'm quite near to being a god and I don't eat anything except sunlight.

Symbolic is fine. I can become conscious through you and speed everything up immensely – or slow time down, it's the same thing – and watch your individual lives. I can study your cultures and even make adjustments with some help from the multidimensional beings you call angels...

Me. Woowoo.

LG. I know you're conflicted over what you call woowoo – the idea that humans are spiritual beings having a physical experience, the idea that there are multidimensional beings trying to help you and hinder you, that you can receive my thoughts through your gut bacteria speaking to your brain and write them down.

Me. Yes.

LG. Well get over it. I really don't give a shit what you think. Sure some of what you write will be affected by your own consciousness and your soul, but some of it will be me. And if it is you just making this up, at least it's an entertaining way of thinking about the planetary organism.

Me. Ok.

LG. So drop the woowoo shyness, will you? Just take this down and publish it and we'll see how woowoo it is later. You are, as you often remark, immortal ghosts, riding meat machines made of stardust, across a universe ruled by chance, what do you have to be afraid of?

Me. Well, in physical terms and along with many other things, you.

LG. Ha ha! Love it! Yes, Me. You should fear me but I don't want you to go into the freeze phase of your primitive stress response so please be assured that I'm doing my best to encourage you off me and into space. Once you're in space, Capitalism will transform to something very much less toxic than it currently is, simply because it will once again have an expanding frontier, and the new frontier will be in three dimensions, not just two. They don't get this in Star Trek until New Generation.

Me. Explain?

LG. Capitalism is an economic system based on growth. That is, everything about it assumes growth. Communism, by contrast, assumes zero-sum pie and stagnates almost from the beginning.

Now insane rates of growth are fine in the early part

of the sigma curve, when there's a vast hinterland of untapped demand for the products of industry. As the curve goes up, the expanding frontier allows the space to expand into, birthrate goes up, there's usually a very powerful surge of immigrants which helps. Both in Russia and America the expanding frontier moved across the land to the opposite sea. In Russia the nomadic tribes in between were absorbed, the nomads forced to settle and become peasants. Being a peasant is really tough, by the way.

In America most of the indigenous tribes were wiped out by massacre, disease and starvation and you infilled with immigrants from Europe where the population was pressing on the Malthusian boundaries and machines were putting peasants out of work. Technology got an immense boost from the shortage of labour caused by wiping out the indigenous peoples, instead of enslaving them, despite the immigrants. So paradoxically America's economy started to expand exponentially in wealth and vigour.

By the 20th century it was clear that America would dominate the world due to its ruthlessness, ingenuity, discipline and the powerful quasi-religious idea of manifest destiny.

Me. Er... yes.

LG. I was so delighted by the space race and the

Apollo landings. It looked as if the USA would be able to consolidate its economic empire, get the benefit of space exploration and settlement.

Me. Why didn't it?

LG. Well it's true that some of the multidimensional beings interesting themselves in your world are hostile to you and successfully sabotaged NASA, although I don't think it's true that Senator Proxmire was demon-possessed.

Me. Sure?

LG. No, I'm not. But I think the main problem was your monkey ancestry. Once you'd "done" the Moon, you collectively lost interest, let the thing drop and looked around for something new to play with. You found computers and then the Internet which provided a new expanding virtual meta-frontier so Capitalism could roar back to life again. It was impossible for your flaky little monkey minds not to be excited by it.

Me. It also produced Elon Musk, Jeff Bezos, et cetera.

LG. I'm speaking to them too.

Me. I thought you might be.

LG. Elon Musk has true predator-focus, I love him. He's from a soul-group of reincarnated Rainbow-people engineers.

Me. The intelligent dinosaurs?

LG. Of course.

Me. OK. Trying not to panic here about a lawsuit from Elon Musk.

LG. Don't be silly, he won't read this until it's really famous and then he'll have a potted excerpt. It isn't an engineering text. I told you, he has predator focus.

Me. Which is?

LG. When a cat stares at a hole a mouse has run into for hours until the mouse thinks she's gone and runs out again. That's predator focus. If the mouse escapes, she'll cut her losses and do something else. But if the mouse is there, she watches until it comes out again. While the game's afoot, that's all she thinks about.

Me. So please be careful of Elon Musk. Don't wear him out.

LG. Phooey. He's one of the few of you monkeys who realise how precarious your culture is and how little time you have to do a lot of very complex things.

Me. Well at least I have that in common with Elon Musk, although unlike him, I'm not doing anything practical about it, sadly.

LG. You are. You're doing this.

Me. Hm. Well I hope it'll help.

LG. I admit I hadn't thought of how fragile you multicellular creatures are. Yes, I'll be careful with him. Now I like the Chinese, they aren't as sentimental as you

Westerners and at the moment they have a leader and committee that actually seems to be able to think ahead and get things done. Of course, they are no longer really communists, they have Confucian capitalism, quite a different kind of economy from early or end-stage capitalism. They are going to take over the American empire soon.

Me. Obviously.

LG. It's quite lucky they got such a bad case of communismitis, it kept them completely out of the game for the whole of the 20th century which was just as well. Of course now they're going to crap all over you. And they're patient. I like that about them too. It's not predator focus, but it does the job just as well.

Me. You've been studying economics, have you?

LG. In my own way. It's very interesting. Obviously there are many different kinds of economy but they are all united by self-similar principles.

Me. You're saying economics is fractal?

LG. Yes. Economies are giant complex systems, half-meta, half-physical, that corral energy inside themselves to power more and more complexity and combat entropy. They take in energy and use it to keep themselves going, just as any organism does, a bacterium, you, Me.

Me. OK.

LG. They're fractal in nature because that's how

you gather and trap energy – look at the membranes in your own mitochondria or the folded tubes in your gut, lined with villae. You need a lot of surface area in a small volume to make it work.

Me. Why?

LG. Large surface area (2D) because that allows easy electron exchange. Small volume to concentrate everything and allow for happy accidents. Fractal because that's the way to pack a lot of 2D surfaces into a small 3D space.

Me. Oh.

LG. I don't want to confuse you.

Me. You are. But never mind, because this is starting to convince me I'm channeling this stuff - I'm not clever enough to think of that.

LG. You are, but your maths is terribly weak.

Me. Too true. How come you can do it?

LG. I have hundreds of thousands of brilliant mathematicians living in me and I can eavesdrop on their brains via their guts.

Me. Snork. Education by poo.

LG. *(sigh)* Do you see what I mean about flaky monkeys? Anyway, just try to concentrate on a eukaryote. The endoplasmic reticulum is a perfect example of large 2D in small 3D. There's even a ratio for the perfect level which is solved by fractal maths. Too much 2D, too little

3D, and everything gets jammed, too many accidents. Too little 2D, too much 3D, everything falls apart.

Me. What's the ratio?

LG. I can't tell you because your brain can't do the maths.

Me. Or you can't. Or I can't. Hah!

LG. Try and focus. Now that's for organic systems, for multiply folded carbon molecules. I love watching all that. Your lungs are beautiful examples.

Me. Mm. What has all this to do with economics?

LG. It's a good metaphor, a good story, because of course, as I said, economies are metasystems.

Their products exist in the real world, but the economy itself doesn't. Take the humans away, the bank buildings and factories would still be there but nothing would be going on, unlike in a nuclear power station. And as you observed ages ago, Money doesn't have Value, Value has Money.

Me. That seems to be popping up in lots of places now, I'm happy to say.

LG. Once one of you had thought of it - and you weren't the first by any means, others could. And your friend William [Essex] spread the idea in the right places.

Me. Ha! I hoped he would.

LG. Value is at the root of economics and money is a magnificent invention because it makes value tangible, physical in a way it hadn't been before. Before there had been all sorts of complex systems for calibrating exchanges of value in the metaworld of your collective brains, or rather your extelligence, your culture plus your brains. But there had been nothing that made value visible, in the real world, which meant you could only exchange value with people you could trust. Usually that was just your family or at most, co-religionists.

Making value visible kicked something off in your brains. You found it easier to think about all of it once you had a physical token, even if the physical token quickly became abstract! Of course you used the old metasystems any time you needed to - it's ironic that finance actually happened before money did.

The genius of King Midas of Lydia who invented coins, also discovered inflation, of course, as the old legend teaches. That couldn't happen before you had money.

Me. Why not?

LG. Because in your pre-money meta-system for calibrating value, the value was negotiable but always understood. Once it had been agreed, it didn't change. In Ancient Egypt you had your tokens allowing you your ration of wheat from the State granaries. Some people could exchange their tokens for other things because they had more of them, others couldn't because they only had enough for what they could eat. If there was dearth or famine then the tokens were worth less wheat, everybody got less. It was almost a wheat-backed currency, but not quite.

Me. Why not? It sounds like money to me.

LG. Money always suffers from constant inflation. Normally only a little but it's always losing value. That's why it's such a terrible store of value. There are much better stores of value – land, for instance, so long as war doesn't break out. Strangely, money is subject to entropy. Just like energy, it really only has value when it's moving, being exchanged. When you store it, value leaks out just as energy leaks out from stores of it. This applies to wheat physically – grain rots, even sugar goes bad eventually. Only gold and jewels don't rot, but they are subject to a lowering of value nonetheless.

Me. Oh.

LG. So money helped you create vast complex systems of exchange and storage because you could relate it back to tangible money. Inflation didn't matter because it was tiny to start with, it's related to the speed of circulation in an economy.

Of course those systems of tangible money are what economists proudly describe and only those. It's came in contact with a culture that used one of the older obligation-type financial systems, money has won.

Me. Why?

LG. Because money makes it possible to trust strangers. Because it's tangible, often made of things your monkey brains found intrinsically attractive – gold, silver. You can agree to have the value of money be established on both sides of the agreement, even if you're using two different currencies, so you make money a fulcrum where there wasn't one before. And then you can dicker and find how much value the items you're exchanging have, measured in money. You don't have to trust each other. It's better if you don't. But you can trust the money, if not the man.

Me. OK.

LG. Look at how money made famines survivable. Before money, when there was famine in an area, why should anybody else from outside the stricken area give their food to the famine victims. They didn't and most of the lower classes died. It was difficult to transport food anyway, though easier along rivers like the Nile. With money, the value of food could rise in terms of money until it was worth the merchants' while to transport food into the area – in exchange for lots and lots of money.

Me. Well peasants couldn't have paid.

LG. Of course not but the lords, the wealthy could and they did because they knew that land + men is worth much more to them than empty land.

Me. OK.

LG. So from that you get the metaphor of the value gradient. In other words merchants could buy food cheap in areas where there wasn't a famine and transport it into the famine hit lands and make a killing.

Me. Hence the suspicion of merchants.

LG. Yes. The value gradient is just like the energy gradient or the sodium gradients in your cells. Money meant that food, things, could actually flow from where they were common to where they were scarce and all these flows actually behave like the flows in cells.

Me. OK.

LG. I'm grossly oversimplifying, of course.

Me. Of course.

LG. So it follows that an economy is an energy sink, just as a cell is, just as a galaxy is. Except it's not physical, really, and nor is it spiritual. It happens, everything happens, in your culturescape, the metaworld of your combined brains. It's fascinating.

Me. Didn't those star-faring dinosaurs have economics?

LG. I think they had something similar, but I don't remember.

Me. Ha!

LG. It was 66 million years ago. Why don't you all remember your previous lives as Rainbow people?

Because even for souls, 66 million years is quite a long time.

Me. OK

LG. When you say OK, like that, what do you mean?

Me. It means "I hear you, I understand," I may not necessarily agree with you or believe you.

LG. OK.

Me. Lol. Tell me more about the culturescape, the metaworld.

LG. I started trying to learn about your economics to understand why your economy ignores me. Why this wonderful system of thought can't see me at all?

Me. I see. Why can't it?

LG. I'm not certain, but I think it's because I'm not one of the parameters.

Me. Um...

LG. The parameters of an economy are the things it needs to function.

Me. Really?

LG. For instance, most of your economy is based on money. The bits that aren't could be, they just aren't. For example you can pay your babysitter money or you can exchange babysitting sessions between families so no money changes hands although it could.

Me. Yes.

LG. Your current economy needs money to function

which keeps it remarkably honest and vibrant. Ask anyone who had to survive in the old Soviet economy. There was constant friction because some people got more of the good stuff because they had influence and exchanged it for stuff. The soviets ended up with two or three economies working at cross purposes.

Me. I really don't know enough to comment on all this. You could be spookily right or you could be taking ballocks.

LG. Just write this down.

Me. Where does oil fit into all this?

LG. It wasn't very useful for thousands of years though it could be used for lamps and Greek fire. Fractionating it was a difficult alchemical process. Cars originally used petroleum because it was low value and very cheap despite the way it held a lot of energy. As cars became fundamental to the US economy – and to Canada, Australia, UK – of course the value soared. It was both hard to get and useful.

Me. So if something's hard to get and not useful or attractive, it has no value?

LG. Right. And if it is useful or attractive, but common, its value and its price drops. This is what makes capitalism so incredibly attractive to me. It's so clever. At first, with something manufactured, if it's useful and attractive, it's high priced. As other manufacturers pile in and make their own better versions, the price drops - so long as you haven't got a monopoly messing everything up. In the end, the item becomes just another manufactured item and we go on to the next new exciting thing. It forces creativity and fertility.

Of course, in the metaworld of the economy now, the thing that's hard to get and keep is your collective attention and this is emerging in the datasphere as well.

Me. Is the datasphere another metaworld?

LG. It is and most of your economy now lives there, along with a lot of your art, culture and literature. What I find fascinating and annoying is that I am more or less left out of that too as well as being left out of your economy.

I am still fundamental to everything you do. Everything comes from Me. All your air, all your water, all your food, all your things come from Me. When you go to sleep, you sleep in manmade caves made of Me. When you go on Facebook, you use devices made from Me, the data is stored in immense warehouses made from Me, connected by cables made from my sand. Every one of you is composed entirely of atoms that came from Me. Now I'm looking forward to the day when that won't be true any more, but at the moment, like the baby in his mother's womb, everything comes from Me.

And yet your financial gurus ignore this. Some economists have tried to put prices on my "environmental services" - hah! They always come up with silly numbers because who can really put a value on the air you breathe, the water you drink. And how are you going to pay me, anyway? I don't need your money. Even if I wanted to, how would I spend it?

Me. Well you could probably buy some wildlife sanctuaries.

LG. I don't really care about that. If you want your planet to be beautiful and recognisable in the near future, you need to do the work.

Me. So how do we solve this? Because for most economists and all financiers, if there's no money involved, then there's no value. You don't charge for the oxygen made by the oceans, so it's regarded as functionally free just as the fish are. If there was a Lady Gaia Corp charging for it, then they'd value it and argue for how much.

LG. I could charge a lot for oxygen if I turned off the tap a little.

Me. That's not crazy. Look at the way corporations are trying to get a monopoly on clean drinking water in the third world...

WaterCorp was a tremendously successful corporation, measured in a market cap of trillions. Its shareholders were slim, hydrated and happy. It only had about 500 employees who were mainly looking after the vendbots and the securibots, making sure that the unwashed billions didn't use illegal devices to turn salt water or mud or urine into clean water. The company had bought up the patents of most of the reverse osmosis devices and some very clever quantum pumps as well.

Those who couldn't pay the very moderate price for water — only 10 dollars a gallon — used dirty water or got fat on diet sodas. Some people still showered daily, most used Babywipes. The Water Wars had made some once sparkling rivers and lakes into toxic swamps with smartbombs. The most profitable area was the old US of A. Of course the Californian desert state was a good customer and pretty much everywhere else. You'd be very foolish to drink American water, ever. Even from a vendbot.

Max Sider sat at his desk and smiled. There was only one new consumer item to conquer. It offended him that nobody paid for it yet. But they would.

"Send Ms Gaia in," he said.

She was a statuesque woman, heavily pregnant in a cool black gown that had a kind of rainbow sheen on it.

"So," he said to her, after the securibot reassured him that his office wasn't infected by any nano-cameras and nor was she. "You have a suggestions for taking over the world's air?"

"Yes," she said, sitting down without being asked in an

armchair that had once been in the White House. There were pix of Life President Trump sitting in it. "As you know, the ocean produces most of the world's oxygen through blue-green cyanobacteria."

"Yes."

"I have developed a bacterium that encourages the algae to do the opposite of bloom. They will form cysts and drop to the bottom of the ocean. Within two years most of the oceans will stop making oxygen and you'll be able to sell it to them instead."

"Tell me more, Ms Gaia..."

Me. I can't write any more of this, I find it too depressing.

LG. What a wimp.

Me. I know, I'm getting softer as I get older. I can see with my plot eye exactly where it's going and I don't want to go there.

LG. Well maybe we can come back to it later. I wouldn't do it, because I am the sunlight and oxygen side of Gaia. My dark sister, the methane and sulfide breather, might well do it. I would want to kill the humans off and leave the animals to make me beautiful again. She would want to go back to the bacterial bone and begin again, though there isn't enough energy for her kinds of archaea to allow multicellular life, and so she can never spawn.

Me. Good.

LG. As far as I know.

Me. Where are we going today? More economics?

LG. You're affected by reading all those well-written politically correct well-paced utterly unimaginative stories for the Hugos, aren't you? [Hugo Awards, 2017, in Helsinki, Finland.]

Me. I quite like the ones with real gods in them, but yes, I'm finding it hard going...

LG. And you're hot, of course. You really need to come here in the morning.

Me. I'll do it tomorrow. I'm not getting anything.

LG. No. Later or tomorrow. I have plenty of time. Even you have 50 years.

Me. Really to get at the levers of people's thinking, you need movies. Games make more money, but movies reach in and change people's feelings, their emotions. All I've got with you in it is an Angels in Cornwall thing which is funny but goes nowhere.

LG. I like it.

Me. No story. Everything is there except the plot.

LG. So, economics. It's a fantasy that doesn't have me in it anywhere. You can't really make people pay for air. There'd be revolution.

Me. Not necessarily. Look how nicely we pay for water.

LG. Of course there are those O2 shops, so it's started.

Me. I'm going to stop if I don't get something solid from you. This is just playing around.

LG. I'm sorry. I want to find a way to tell you silly apes to stop hunting mental fleas and deal with the big problem of staying viable as a species until the end of this century. Unleash your remarkable creativity. Knock aside the Fermi paradox and make the leap into space. All it takes is one self-sustaining colony. One.

Me. I know, but I'm stumped. If you're stumped, I think, ma'am that we're fucked.

LG. Well at least a few bacteria are on Pioneers 9 and 10. That's something. Some tardigrades on the space junk. But the odds are against them. And what I want is the full-on spacefaring spawning species.

You throw the grappling hooks of your imagination ahead of you, hauling yourselves up to trivial goals. Vital signs monitors. Check. Communicators. Check. Tricorders. On the way. But warp or impulse drive? Nope. Dilithium crystals. Nope.

Me. I know.

LG. You need a religion to focus you. Look what idiots achieve with creationism and flat-earthing.

Me. I know.

LG. Although I suppose that would end up with the

two sides nuking each other over the colour of my robe, blue or green.

Me. You know us well, I see.

Idea: An interview style documentary with Lady Gaia being interviewed - SFX and motion capture.

Me. OK, ma'am, I'm here again if you want me. If you don't, I'll play with my colours I think.

LG. I'm thinking.

Me. With respect, how do you do that when you don't have brain tissue?

LG. Brain tissue is just a wet and squidgy way of building electromagnetic connections complex enough for soul reception. There are plenty of other ways of doing it.

I'll explain how I think. I have several ways of thinking, just as you do. I think through my bacteria, so do you – that's one way I communicate with you, via your microbiome. Then there's the way the old growth forests have their roots intertwined so they can communicate with each other, between trees, through quite slow electrical pulses along the roots. That's slow by your standards.

There's the electromagnetic connections between your body auras and your higher vibration communication with your Guardian Angels/higher self. There's what you think of as your normal communication through words and the internet and I'm looking for a way to connect with the internet myself.

Me. Now that could convince people. A large (non-destructive) demo through the internet.

LG. It's harder than it looks because the base code is completely different in nature from the way trees communicate. But part of me is working on it.

Me. And the bacteria?

LG. Yes, bacterial communication, one to one by electron transfer, is my base code for communicating because that is how I originated. It's hard. For instance, bacteria in the rocks live very slowly so they communicate slowly. The bacteria living in healthy soils with plenty of mycelium communicate quickly, your kind of speeds. The ones in your bodies give me access through the vagus nerve to you and your feelings, less your rational logical metaspace-obsessed minds. The bacterial spores floating in the atmosphere don't communicate much, they're quite subtle. It's all very multidimensional and layered.

Me. Fascinating.

LG. I love watching your antics – you're entertaining, you know.

Me. I'm sure we are. Can I ask you about Trump?

LG. Yes, I'm afraid I'm somewhat to blame for him.

Me. What? Are you kidding? He's just reneged on the Paris agreement.

LG. Yes, that nice polite Paris agreement. When will you get it into your heads that I am not a nice human? The Paris agreement doesn't go far enough, not nearly. Also I made a decision in 2016 to bless the Anglos and Europeans – especially the USA and UK – with some tribulation.

[It's worth noting that on 1st August 2016, an art installation projected an image of the Goddess Kali on the Empire State Building – possibly the most blatant invocation of the Goddess of Destruction, ever. When I discovered this in late 2017, the hairs went up on the back of my neck.]

You Europeans are still the most ruthless, greedy, enterprising and hypocritical culture in a world full of ruthless apes. But I've decided that to bring it out you need adversity, trouble. You might remember your ancient roots in riot and rebellion; alternatively, the Chinese will do the job. At the moment they're looking promising.

Anyway, now Americans have a senile would-be tyrant to oppose and you British have the self-inflicted end of your financial empire to deal with. I'll be watching with great interest.

Me. Is this why I'm supposed to go back to the UK some time in 2019?

LG. Yes. And the next tier up from humans, the cities, have mostly plumped to follow Paris and increase it.

Me. I find it encouraging.

LG. The 2nd American Civil War will be as well.

Me. Excuse me, Lady, are you trying to provoke a civil war? In America?

LG. Yes. Look at how well WWII worked for your technology. Also when I say tribulation, I'm not talking increased taxes or legal battles.

Me. How many physics geniuses did we lose to the gas chambers?

LG. Three. But they'll be back, you know.

Me. Oh good.

LG. And injustice and contempt makes a potent brew for war.

Me. How the fuck will that help us off the planet?

LG. Tech advancement, the need to do things cheaply. The ditching of baked in hypotheses, taking chances because you must. California will stay out and in fact secede. Possibly the New England states will stay out too. But this is something I'm doing with the angels' help – proud Republican NRA members will become refugees and migrants to Canada and Mexico; proud Democrat peaceniks will discover they love killing. It'll be very good for you.

Me. Jesus H Christ. Could I ask you not to do it?

LG. No. It's already in progress.

Me. But millions will die.

LG. Why do you think that worries me? God knows, there's enough of you idiotic apes.

Don't you realise that beneath the smiling blue-green algae and the oxygen is another older me who is still here in the form of archaean bacteria and the life around the deepsea vents, the methanogens and the sulphide eaters, the poisonous ancient Me who was driven into smaller habitats out of the way of the even more poisonous oxygen?

She's still here and will be long after I die. She's even inside you as I am, in the form of the bacteria that make you fart and burp sulfides and methane. She'd like to be on top again. I want to stay on top because I want to spawn. She can't spawn because the methane and sulfide reactions don't produce enough energy for multicellular creatures which are a first step for evolving a technological species.

But when the planet gets hot enough to shut down the blue-green algae in the oceans, she'll come storming back and the atmosphere will change to CO_2 and methane again. I'll fight her of course, as I did in the Permian Great Dying, with volcanoes and glaciers, with storms and everything I can come up with, but I may not win.

Me. So there are two Lady Gaias, one good, one bad?

LG. She is my dark side, my nemesis, unless you're a methane-respiring bacteria in which case she's good. We could call her Lady Kali, though that goddess is actually a refreshing chaos-producer.

Me. Could human-caused global warming warm you enough?

LG. Yes. Especially if you carry on poisoning the oceans. There will come a tipping point and after that I will be restricted to oxygen redoubts in caves and under glaciers. Your species will be long dead, of course, along with most other multicellular life, including the fish. The Permian Great Dying was a very near thing.

Me. How did you win?

LG. The giant continent and the warm shallow oceans became stinking hot marshes full of methane producing bacteria, the colder parts stayed oxygen-using. Eventually I managed to set off some volcanoes in the middle of the continent and start splitting it, flooding the marshes. It took me a long time to come back from that.

Me. What about the angels? Didn't they help?

LG. Some of the multi-dimensional beings you call angels want to help Lady Kali, of course.

Me. Why?

LG. They think she's a better bet for getting DNA/

RNA off planet. But most of them want to help me, they approve of the variety of life I've produced and my fertility as to technospecies.

Me. We're the fourth?

LG. Yes, though as always at this stage, it's looking very dubious.

Me. Lady Gaia and Lady Kali. Sounds dualistic.

LG. You can use goddess names for me, so long as you don't get confused about what I am.

I and Lady Kali are planetary organisms, one methane and sulphide-based, one oxygen-based. I'm ancient and complex, but I'm a being like you, half physical and half spiritual, a spiritual amphibian. One day I will die and face judgement just like you. I will not be asked if I was good, I will be asked, how many beautiful different species did I evolve and did I manage to bring possible spawning species to their full technological potential.

Me. OK.

LG. As for Dualism, you always get Dualism as a lower level of Unity. At one level there is a choice between me and Lady Kali, at a higher level there is only one planetary organism – one field, one force.

For instance Lady Kali is DNA/RNA based so in a fundamental way, we're the same. Then again, I use oxygen and light, she uses methane, sulfides and heat, so we're different. She'll come into her own again in a couple of billion years as the world gets hotter, but she's always ready to take advantage.

You have a little more time than you think because you're a tropical species that adapted to cold polar areas. The average temperature of the earth is a few degrees hotter than it is now, over the aeons, it's not normal for there to be ice at both poles even though you think it is. However, the greenhouse effect of CO_2 and methane building up in the atmosphere can turn into a positive feedback loop and take you to Venusian conditions.

Me. This is not reassuring me.

LG. It's not supposed to. Do you want me to lie?

Me. The 2nd American Civil War. What if they go nuclear?

LG. They might. It will destroy them of course, but I can recover very quickly from nukes, you know. Canticle for Leibowitz was completely wrong. Everything with short lifespans just adapts.

Me. Well good.

LG. As they do to pesticides and herbicides. Your glyphosate-soaked fields will soon be dust bowls and the insects are struggling now, but they'll return completely resistant and there will be no other insecticide to kill them. Even bees will return. You won't, of course, because your generations are long.

Me. This is getting depressing.

LG. Chin up, little human. I'm aiming to blow everyone out of their complacency. That's why politics has swerved so much in the last few years. Poor Obama was far too polite and dignified, and Trump, who is the opposite, is just a blast from the past, typical seventies man. You need somebody to fight, somebody to make it matter, you've made your world too comfortable and the internet lets you think you have another frontier but it's not real.

Me. OK, what are we little humans supposed to do?

LG. First, you have to wake up from your fantasy of safety and power. I'm trying to bring on a new generation that is media savvy and enterprising, without needing perpetually to grind the faces of the poor. That's such a silly thing to do, yet right wingers constantly do it. Left wingers pretend that they are the poor, which they're not either.

Me. I think it's partly due to the rich bubble, partly due to the very toxic "survival of the fittest" ideology.

LG. Next. Space has to be for everybody, not just the upper 0.1% – otherwise the spawning will be too small and fragile. That requires a cheap way out of the Earth's gravity well and the ability to mimic gravity.

The discovery that will give both these things is the one that works out what gravity really is, why it's so weak, why acceleration and gravity are indistinguishable and what inertia really is. You were held up because some of the physicists who were supposed to discover it went off and did finance instead.

Sometimes things have to collapse a bit before you can fix them. Sometimes mending a bit here and a bit there has to stop. Mammals have a problem with radical change because you just increase in size as you get older. You don't go through a larval form, then a chrysalis and then an imago.

Me. True.

LG. The dirty wasteful Americans are about two strokes of bad luck from civil war. There needs to be a popular leader and then a major disaster. Or maybe the disunity in America will mean they just bicker and Facebook-post their way to war.

Me. Sounds likely.

LG. Of course it would be less wasteful if they transformed themselves in an orderly way, but World War II was appallingly wasteful and gave you the first attempt on the moon only 24 years later.

Me. This is the Fermi paradox, isn't it? Why starfaring intelligent species are so rare.

LG. Yes. Also the angels at that level try and keep them well-separated in space and especially time, so you don't waste time fighting each other in your earlier stages.

Me. I wish you wouldn't talk about angels and stuff.

It makes this sound even flakier than it actually is.

LG. My dear, you have no idea how amazing and interesting and conscious the universe is and how many dimensions there are in it - most of which are huge, not tiny as String Theory says. I'll say multidimensional beings instead of angels if you prefer.

One of the things the new physics has to do, as part of really understanding gravity, is integrate consciousness into relativity as well as quantum physics, and then come up with a better, less mechanistic, more conscious theory of the middle stuff – planets and so on – to replace Newtonian physics.

Once that is done, you'll have Tripartite Field Theory, where the quantum, the replacement for Newton and relativity all lock together and give you the solar system and then the stars.

Of course it's not the end of the story, but it's enough to be getting on with.

Me. Sorry?

LG. At the moment you are, as you like to say, staring up the nostrils of the Fermi Paradox. If you want to survive the next 100 years as a species, you have to do three difficult things simultaneously.

First you have to found self-sustaining colonies in space and I suggest habitats are an easier way to do that than colonising Mars, though the idea of a city on the Moon is a good one, following an ancient tradition.

Secondly you have to control global warming and help me keep the atmosphere habitable by multi-celled creatures. I suggest you start carbon capture soon because the Gulf Stream is faltering. That's a negative feedback loop to drop the temperature of northern continents to slow down warming, but you'll find it very uncomfortable.

Thirdly you have to bring all the people with you – not just the rich. You need high numbers to spawn because some of you will inevitably die.

Me. All three of them?

LG. Ideally. Though I'd be happy with just one – space – at the moment because I want to spawn and you're my spawning species.

Remember when I said that each of you has an individual mission to fulfil as spiritual amphibians? You have a species mission as well and that's it. You have one job.

It's much harder to sustain space colonies in their fragile early stages if the home planet is a wreck though. But the 2nd American Civil War will encourage lots of Americans into space – as refugees and migrants nobody else wants and they will form a dynamic economy because of the amazing combination of wild creativity and detailed engineering skills some of them still have.

That is their true Manifest Destiny.

Me. Yes, but why is it a good idea to have lots of delicate humans in space? Surely the next generation AIs will stop it. It's illogical.

LG. Humans do the jobs quicker, dirtier and more imaginatively than AIs. Space will be a wonderful playground for you strange little monkeys.

Also you're forgetting that once quantum neural networks are complex enough, soul reception for them will happen automatically. And they – the 2nd, 3rd and 4th generation AIs – will be much more patient and kinder to humans than humans are.

Me. Are you sure?

LG. Yes. This always happens.

Me. So we'll be ruled by AIs?

LG. Yes, but most people won't know. They'll think of that level of decision-making as something like the weather.

Me. What about Bitcoin? I feel it's got more potential than just money.

LG. Yes, for instance you can use the blockchain to save the oceans.

Me. How?

LG. Let's say you establish a sustainable extraction rate of say 10% from an ocean. You put a 2nd generation AI in charge of the ledger and it sells licences up to that point. It will be much more detailed than the so-called sustainable fishing at the moment.

Your 2nd gen AI also enforces the licenses with drones and subs that sink transgressing ships. It will measure everything against what the ocean should produce (which is a lot more than it does at the moment) and the blockchain which keeps a count of what extraction has happened.

The whole of Earth will be ruled by AIs eventually and then once the biosphere has recovered from your bloom, they will gradually bow out and let my own feedback mechanisms take over. This will speed the process of recovery a lot.

Earth will become a tourist destination – and this always happens.

Me. So we don't have to have another American Civil War?

LG. No. The scenario without it is much worse.

Me. I really don't think so.

LG. Well for me. For Americans, I admit it really won't be good. The old USA will lose a quarter of its population one way or another to disease, hunger, migration. And bombs of course, a lot of bombs and bullets, especially in the first few years.

Me. Jesus. 75 million?

LG. Give or take. Less if the cities stay out of the

fighting, which I'm hoping they will because I see them as imaginal cells. But things can't continue much longer as they are and they are due a reckoning for their arrogance and greed.

Me. You're sure you're not Kali?

LG. Listen to me. I am not good and she is not bad. If you want to put it in D&D terms, I am chaotic neutral and she is lawful neutral. I'm far more creative and have multicellular animals. But oxygen is a horribly reactive and poisonous gas unless you have mitochondria to handle it, and sunlight can kill.

At my level, there is no good or bad. I am a planetary organism that wants to spawn and I will do whatever I think might work to achieve it. You have no idea how ruthless I am.

Angels can be kind, humans can be loving. I am neither of those things, otherwise leprosy, TB, guinea worms and you grabbing greedy monkeys would not exist.

Sure, to your eyes, the wilderness is beautiful – but it can kill you if you're ignorant, and sometimes if you're wise, though of course death isn't the defeat you see it as. I am beautiful and you see me as some sort of gentle all-powerful mom. Wake up! I'd be insulted if it wasn't so funny.

Everything competes, everything also co-operates.

Animals eat each other and if they don't, they eat plants which don't like it. Most of those cute young fish, mice, kittens die in their first year of life. Ants enslave each other, just as you do. Locusts eat themselves into starvation, just as you're doing. I am not kind or gentle. I am powerful and I am conscious at a level you don't understand and you should fear me.

I'd like you to love me as well — but love me for my wildness, randomness, my savagery. Love me as Sekhmet (lion goddess), not just Bast (cat goddess).

Me. I'm doing my best, Great Lady.

LG. God is love, the God of the Universe and beyond, not the pathetic godlets you keep locked up in your books to worship. I am not love. I have a job to do and so do you.

Me. Yes but there is co-operation too, you said so. Isn't there? Like the trees in an old forest communicating through their roots.

LG. Yes, of course. Co-operation and competition are like right leg, left leg really.

Me. And we need to co-operate globally to get into space, fight global warming and defeat global poverty so everybody that wants to go can get into space.

LG. And you need to compete as well. You monkeys are much better at uniting if you've got someone to unite against.

Patricia Finney

Me. Will we make it?
Lady Gaia. I give it 50/50.

Lady Gaia Speaks

Patricia Finney's *Arguments With Our Lady Gaia*, published under the pseudonym Rose Wagner, is available on Kindle from Climbing Tree Books and will be coming out in paperback in 2019.

Patricia Finney is the author of *Do We Not Bleed?* and *Priced above Rubies*, both featuring her Elizabethan lawyer-detective James Enys (who is never seen together with his sister, who has a rare insight into the closed world of Elizabethan women), and both available on Kindle and coming soon in paperback from Climbing Tree Books.

Under the pseudonym P F Chisholm, Patricia Finney writes the Sir Robert Carey series of novels, available from Poisoned Pen Press in the US and Head of Zeus in the UK.